Orbits

the energy generators

By

Keith Dixon-Roche

ORBITS

All concepts and formulas in this book not previously attributed to Johannes Kepler and Isaac Newton, are the sole property of Keith Dixon-Roche and protected by copyright.

Their use, publication, broadcasting, distribution, copying or any form of recording without Keith Dixon-Roche's written consent shall be a breach of international copyright law and subject to legal action.

Copyright © Keith Dixon-Roche 2017 to 2026

Orbits

the energy generators

Published by CalQlata

info@CalQlata.com

First published October 2024
Final publication April 2026

This book is sold subject to condition
that it shall not by way of trade or otherwise,
be lent, re-sold, hired out or otherwise circulated
without the publisher's prior consent and in such
circumstances it shall not be circulated in any form of
binding or cover other than that in which it is published
Copyright © CalQlata 2017 to 2026

ORBITS

Contents

Preface 1

1 Introduction 3

2 Energy 5
2.1 Kinetic Energy 6
2.2 Potential Energy 7
2.3 Distance & Time 8
2.4 Charge (electrical & magnetic) 9
  2.4.1 Magnetic Energy 9
  2.4.2 Electrical Energy 9

3 Orbits 11
3.1 Orbital Laws 12
3.2 Elliptical Orbits 13
3.3 Circular Orbits 14
3.4 Linear Orbits 15
3.5 Centrifugal Force 16
3.6 Station Keeping 17
3.7 Orbital Precession 18
3.8 Orbital Planes 19
3.9 How They Work 20
  3.9.1 Newton's Laws of Orbital Motion 20
  3.9.2 Elliptical Orbital Formulas 22
  3.9.3 Circular Orbital Formulas 24
  3.9.4 Linear Orbital Formulas 26
  3.9.5 Centrifugal Force 27
  3.9.6 Station-Keeping 28
  3.9.7 Orbital Precession 29
  3.9.8 Orbital Planes 30

4 Universal Orbits 31
4.1 The Great Attractor 32
4.2 The Milky Way 33

4.3	Hades	34
4.4	Our Solar System	36
4.5	Our Sun	37
4.6	Our Solar System	38
A-1	*Glossary*	*41*
A-1.1	Terminology	43
A-2	*References*	*45*
A-3	*Useful Formulas*	*47*
A-4	*Newton's Orbital Laws*	*49*
A-4.1	Nicolaus Copernicus (1473 to 1543)	49
A-4.2	Johannes Kepler (1571 to 1630)	49
A-4.3	Galilei Galileo (1564 to 1642)	49
A-4.4	Isaac Newton (1642 to 1727)	51
A-4.5	Proof (elliptical orbits)	53
A-4.6	Euclidean Geometry (equal areas)	54
	A-4.6.1 Proof (conservation of energy & equal time-swept area)	55
	A-4.6.2 Centripetal Force	58
	A-4.6.3 Distance Between a Satellite & its Force-Centre (R)	59
	A-4.6.4 The Inverse Square Law	60
	A-4.6.5 Orbital Period	61
	A-4.6.6 Constant of Proportionality	62
	A-4.6.7 Alternative Velocity Calculation	63
	A-4.6.8 Centrifugal force in an orbiting body	64
	A-4.6.9 Fundamental Laws of Orbital Motion	65

Preface

Johannes Kepler discovered the laws of orbital motion and Isaac Newton completed them; Copernicus, Galileo, Poincaré, Huygens and Halley also lent a hand.

Having studied various publications on this subject, I concluded that we haven't actually understood them all that well. For example, how we deal with;

>planetary spin
>
>binary stars
>
>centrifugal force in non-circular elliptical orbits
>
>station keeping
>
>core pressure
>
>galactic force-centres
>
>orbital precession
>
>orbital planes
>
>circular orbits ($E=mc^2$)
>
>etc.

none of which have been adequately explained.

I have always believed that unless every aspect of an event or characteristic can be explained *from first principles*, it will never be properly understood or fully exploited; and because orbits are the answer to all of our needs, why not try to understand them better.

Our lack of understanding of orbits is the reason we today need to extract neutron energy from radioactive matter, and why we spent billions of dollars and millions of manhours to create the atom bomb, which could have been understood in days and worked out on the back of an envelope if we had understood the workings of the atom. It is also the reason why we've built colliders (Hadron and JET) to search for things that don't exist and projects that can never succeed.

So, I decided to try and solve the missing details myself. Once I fully understood orbits, I realised that Relativity not only fails to work, it is actually unnecessary. Moreover, having discovered that they also apply to atoms, and the true meaning of Poincaré's relationship; $E = mc^2$, I realised that they are the source of all universal energy.

Keith Dixon-Roche

2024

ORBITS

1 Introduction

Orbits – circular, elliptical and linear - are universally essential as they are responsible for generating all of its energy; atomic and celestial.

This publication explains how they work, including their mathematical formulas and applications; atomic (circular), celestial (elliptical) and universal (linear).

The energy released during the last 'Big-Bang' was induced as kinetic energy into the masses ejected from the ultimate body, and when acting together with the potential energy (gravity) between these masses, resulted in orbital structures; galaxies. The magnitude of this energy release is the total energy in all universal orbits; atomic and celestial, today, and will remain as such until the next 'Big-Bang'.

As long as outer-space is a perfect vacuum (forgetting dark matter), these orbital systems will remain in-tact until the next 'Big-Bang'. And the re-accretion of all universal matter at the end of this universal period will result in a new ultimate body, which will in turn generate a new 'Big-Bang'. This process can continue into infinity with no outside help. It is this re-accretion and sequence of 'Big-Bangs' and universal periods that were responsible for the dispersal of the life-making proteins we find in all celestial matter.

ORBITS

2 Energy

Our universe comprises electrical and magnetic energy and nothing else; there is no such thing as mass or gravity.

Mass is a term of convenience for atomic particles - and collections thereof - that was derived for an unknown property, and gravity is the magnetic attraction between them.

Energy was not a concept known to Isaac Newton, so he used force to describe energy transfer; which is the manifestation of energy transferred between two or more bodies separated by physical distance.

$$F = G.m_1.m_2 / R^2$$

therefore;

$$E = G.m_1.m_2 / R$$

The entire universe comprises a fixed, unchanging quantity of energy, it always has done and always will. It was originally contained within the neutrons in the ultimate-body, released during the latest 'Big-Bang'; it remains unchanged today.
[first law of thermodynamics]

The energy released during the last 'Big-Bang' was ≈7.4E+60 Joules, which became the total energy (E) in all universal orbits. And because planetary spin (heat) is generated by orbits, and because energy cannot be created or lost - it can only be transferred - all the energy in the universe must be ≈7.4E+60 Joules.

The two types of energy that dominate orbits are potential and kinetic.

Because a satellite's orbital positive KE and its negative PE vary throughout an orbital cycle, total energy always (E = KE + -PE) remains constant. Also, because of the conservation of energy and the fact that the orbit is a symmetrical ellipse, the potential & kinetic energies on both sides of its orbital principal axis are equal and opposite.

ORBITS

2.1 Kinetic Energy

Kinetic [dynamic] energy (KE), which exists in all moving particles, is always positive and induced via electrical, magnetic (gravitational), electro-magnetic or impact (potential energy).

Kinetic energy in a satellite following a circular orbit (such as in an atom) is not induced into the satellite by its force-centre such as in elliptical orbits; it must be provided by the satellite itself.

Its linear (or curvilinear) mathematical relationship is:

$$KE = \tfrac{1}{2}.m.v^2$$

The kinetic energy of a satellite orbiting in a circular path is exactly half the satellite's potential energy:

$$KE = \tfrac{1}{2}.PE$$

Kinetic energy can only be converted to kinetic force if the moving satellite is restrained by magnetism (gravity) or a physical connection dependent upon their separation distance (R), and it may be calculated for elliptical orbits like this:

$$KF = KE / R$$

where 'R' is the satellite's orbital radius

and because this force acts in a straight line between the satellite and its force-centre, it becomes a potential force. But this is not the same force as that induced by the magnetic (gravitational) attraction between the satellite and its force-centre.

And the potential acceleration (g) induced by a moving body may be calculated thus:

$$g = v.v^A / R(1-e) \qquad KF = m.a$$

where v^A is the satellite velocity at its apogee and m is its force-centre mass

2.2 Potential Energy

Potential [static] energy (PE), is the positive or negative, magnetic or electrical attraction (or repulsion) between a satellite and its force-centre. It exists between all Quanta in the universe, irrespective of separation distance (d).

What we currently refer to as gravity is actually the magnetic potential energy between all universal Quanta.

> Negative potential energy, e.g. gravity, holds particles together, and;

> positive potential energy, e.g. centrifugal (CE), drives them apart.

Whilst the potential energy radiated by both magnetic and electrical charge retain their magnitude irrespective of distance, it is distributed over the spherical area at that distance. This is why such a potential force between any two bodies (gravity) *appears* to diminish with the square of the distance between them.

Its linear mathematical relationship is:

$PE = m.a.d$
in which 'a' may be positive or negative.

In a balanced system (e.g. orbits), both PE and CE must be equal. And in circular orbits, such as atomic, potential energy between Quanta is always twice the kinetic energy in the orbiting satellite, so their mathematical relationship is:

$KE = \frac{1}{2}.m_2.v^2$
in elliptical orbits;
$$g = v^A.v^P / d.(1-e)$$
in circular orbits; $v^A = v^P$ & $e = 0$, therefore;
$$a = v^2/d$$
$$PE = m_2.a.d = m_2.(v^2/d).d = m_2.v^2, \text{ therefore;}$$
$PE = 2.KE$

Potential energy may be converted to potential force by dividing potential energy by the distance between the bodies. It may be calculated like this:

$PF = m.a$

which is the same force described by Isaac Newton and Charles Augustine de Coulomb:

$F = G.m_1.m_2 / d^2$

$F = k.e^2/d^2$

and which refers to the magnetic (gravitational) and electrical attraction between a satellite and its force-centre; assuming the electrical poles of the force-centre and its satellite are equal and opposite.

2.3 Distance & Time

In physics, distance (d) is the length of path between two points; either straight or circuitous. And time (t) is the passage of time between events.

Distance and time are two of the only four variables required to explain all the properties of every branch of science and engineering; energy.

Neither universal time or distance deform around massive celestial bodies, they are both universal constants;

> A unit of distance is always exactly the same, anywhere and everywhere throughout the universe
>
> A unit of time is always exactly the same, anywhere and everywhere throughout the universe

Distance and time are used to define velocity ($v = d/t$) and acceleration ($a = d/t^2$)

ORBITS

2.4 Charge (electrical & magnetic)

Every electron possesses exactly the same electrical (e) and magnetic (m) charge (mass) of exactly the same magnitude, everywhere in the universe.

Every lone proton possesses exactly the same electrical (e) and magnetic (m) charge (mass) of exactly the same magnitude, everywhere in the universe. But as part of a proton-electron pair, it will collect and hold onto, an additional electrical charge (e') that is transferred from its electron partner dependent upon the electron's orbital kinetic energy.

Along with distance and time, electrical and magnetic charges, are responsible for all universal energy.

2.4.1 Magnetic Energy

Newton's magnetic [charge] energy is calculated like this;
$$E_M = G.m_1.m_2/d$$
and because magnetism is accrued, m_1 & m_2 may be any value in his formula.
Magnetic energy may also be calculated like this;
$$E_M = m.(d/t)^2 = m.a.d$$
This is the reason for such huge gravitational potential energy between celestial bodies.

2.4.2 Electrical Energy

Coulomb's electrical [charge] energy is calculated like this;
$$E_E = k.e_1.e_2/d$$
but because electricity is shared, e_1 & e_2 must be equal in his formula, i.e.;
$$E_E = k.e^2/d$$
where 'e' must be the least magnitude of the two (e_1 & e_2).
This energy may also be calculated like this;
$$E_E = e.(d/t)^2$$
This is the reason for negligible electrical potential energy between celestial bodies.

Important Note:
These charge energies are radiated by - and act between - every atomic particle throughout the entire universe. Whilst it *appears* to diminish with the square of the separation distance, it doesn't, because this would contradict the conservation of energy. This apparent diminution is due to the fact that these energies are distributed over the spherical area at that distance ($A = 4\pi.d^2$).

ORBITS

3 Orbits

Orbits are ubiquitous; they are everywhere in the universe, both celestial and atomic.

All universal energy is generated by planetary spin (and fission) in all the bright stars and planets. It is then stored in the neutrons created in the same bright stars and planets.

All the energy in the universe ($\approx$7.354E+60 Joules) originated from the neutrons split during the last 'Big-Bang'. This is the energy driving all universal orbits.

Orbital energy is constant and eternal (in a vacuum), and the source of internal heat in celestial bodies through planetary spin, which is the instrument of neutron creation through atomic fission in bright stars and planets.

The heat we feel on the surface of our planet from stellar radiation is the EME generated by the proton-electron pairs in the sun's atoms.

There are no such things as binary stars. Those which today's astronomers claim are binary stars comprise one bright star, orbited by a bright planet. Like bright stars, bright planets have collected sufficient sub-satellite mass to create the heat required to generate fissionable energy within its core atoms.

A force-centre/satellite partnership will remain until the satellite is physically removed. Even satellite rubble that has been decimated through impact will remain in the same orbit after destruction; which is the origin of asteroid belts.

The force-centre is always located at the focal point of the orbital system, which is on the major axis and the closest point at which an orbiting satellite passes by.

A satellite's orbital motion is maintained by the increasing potential energy (PE) between it and its force-centre raising its kinetic energy as it travels towards its perigee, and subsequent decrease as it travels towards its apogee. *In a perfect vacuum*, this motion is perpetual; it will endure indefinitely. This is how we know that there is no such thing as dark matter.

ORBITS

3.1 Orbital Laws

The following fundamental laws apply to every natural orbit, without exception:

1) Every orbital system has only one force-centre and at least one satellite.

2) Altering the properties of a satellite will not affect its orbital shape or period.

3) Every orbital path is an *exact* ellipse, the eccentricity of which is; $0 \leq e \leq 1$

4) Every orbit about the same force-centre has an identical constant of proportionality (K).

5) The constant of motion (h) of every satellite applies throughout an elliptical orbit.

6) The total energy in a satellite (E = PE+KE) is a constant throughout its orbit.

7) Spin plays no part in Isaac Newton's mathematical laws of orbits.

8) Once united, an orbital partnership will remain intact until or unless it is physically broken.

9) In circular orbits (e = 0), a satellite must provide its own kinetic energy, which is always exactly half the potential energy between it and its force-centre;

$$PE = 2.KE = 2 \cdot \tfrac{1}{2}.m.v^2 = m.v^2$$

ORBITS

3.2 Elliptical Orbits

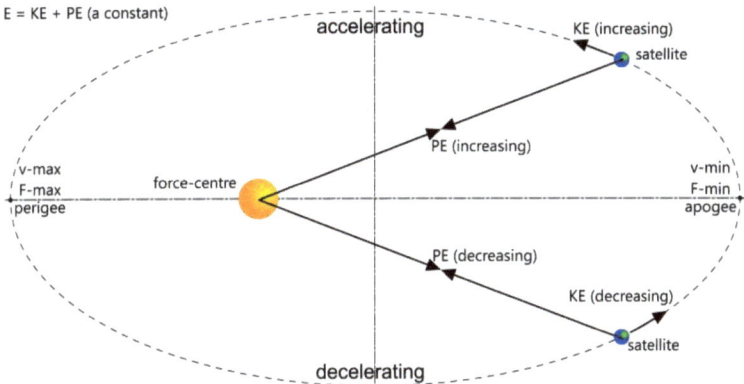

Non-circular elliptical orbits are those with an eccentricity; $0 < e < 1$

These orbits are self-generating in that the variable potential energy in a satellite induces in it a variable kinetic energy, which reach a maximum and minimum at its perigee and apogee respectively.

As such, this orbit will continue unchanging until the next 'Big-Bang', but only in a vacuum. In other words, only if there are no atmospheric particles (e.g. dark matter) to absorb the satellite's kinetic energy.

In celestial bodies, the competing energies (potential and kinetic) in a force-centre, its satellites and sub-satellites are responsible for generating internal frictional heat within a satellite through planetary spin.

3.3 Circular Orbits

Circular orbits are those with an eccentricity; e = 0

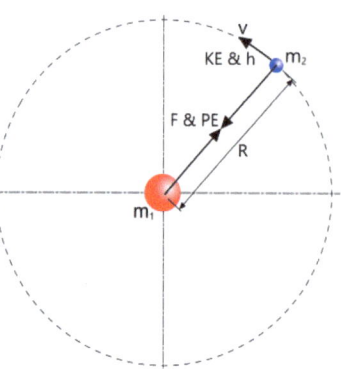

These orbits are not self-generating, in that the satellite must provide its own kinetic energy. They refer specifically to man-made geo-political satellites and proton-electron pairs, and are the justification for Henri Poincaré's formula; $E = mc^2$

Whilst these are calculated using exactly the same formulas as elliptical orbits, their circular nature makes these calculations much simpler.

ORBITS

3.4 Linear Orbits

Linear orbits are unknown today, because the origins and workings of our universe are unknown. However, they are real, and obey the same rules of motion as elliptical orbits.

The only differences between linear and elliptical orbits are that a linear orbit …

1) is a single event
2) generates no internal frictional heat in its force-centre
3) has no constant of motion (h)

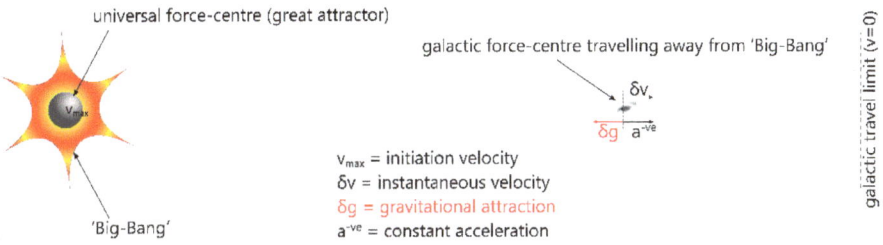

However, a linear orbit does have a constant of proportionality (K).

In reality, our universe is <u>now</u> understood (see Appendix A-2), which means that linear orbits are <u>now</u> known. All galactic force-centres are travelling away from our universal force-centre at similar velocities that vary according to their galactic force-centre masses.

Because the galactic force-centres are travelling in a straight-line, no internal frictional heat is generated within the universal force-centre (the great attractor).

Whilst our universal linear orbits are single events, they self-repeat automatically after every 'Big-Bang'. The great attractor is the re-accreted rubble left behind following a 'Big-Bang'. This body, is our universal force-centre.

Each galactic force-centre is a [linear] satellite and its stars are its sub-satellites. Planets are sub-sub satellites, and moons are sub-sub-sub satellites. The gravitational pull from the great attractor will eventually stop outward travel in all of its satellites (galaxies), and pull them back into itself. When the re-accreted mass is sufficient to compromise its innermost neutrons, a new 'Big-Bang' will occur.

3.5 Centrifugal Force

Centrifugal force in a satellite is the potential force opposing its force-centre's gravitational force. A satellite will only remain in orbit as long as the centrifugal and gravitational forces are equal (and opposite); *station keeping*.

For any given satellite, centrifugal force has one constant (mass) and one variable; (acceleration).

Whilst the calculation for **centrifugal acceleration** is relatively straightforward in circular orbits;

$$F_c = m \cdot v^2 / R$$

As you will see in chapter 3.9.5, it isn't quite so straightforward for elliptical orbits ...

3.6 Station Keeping

When an external displacement force attracts a satellite, attempting to pull it off course, a restoring force will keep it in its orbital path, maintaining balance in its centrifugal and potential accelerations; i.e. keeping these forces equal and opposite.

This relationship is what maintains a satellite's orbital path. Even if a satellite is destroyed through impact, the resultant rubble will remain in the pre-impact orbit.

This process requires a perfect elliptical orbit; exactly as Kepler explained. It does not work with Relativity.

3.7 Orbital Precession

Kepler believed that satellites exposed to the gravitational influence of neighbouring satellites would be temporarily pulled out of orbit. But because station-keeping maintains a satellite's orbital shape, the effect of gravitational attraction by external bodies is to induce a torque on a satellite's orbital axes, causing them to rotate about the orbital force-centre.

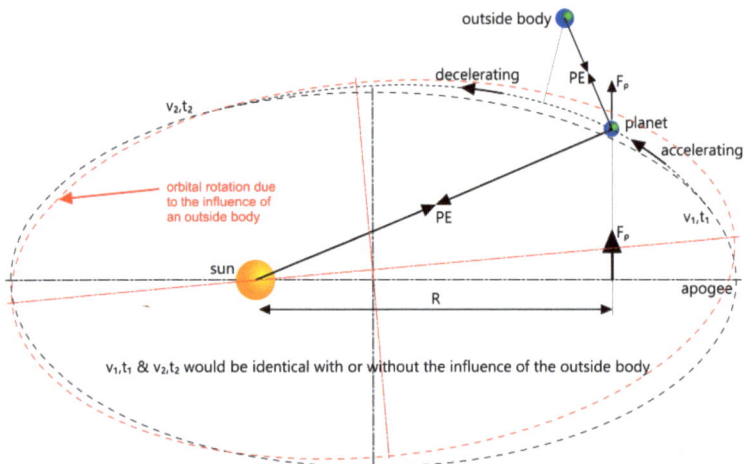

This orbital rotation is the reason celestial satellites in adjacent orbits occasionally impact, creating galactic and solar comets, which occasionally become trapped by other satellites; creating new planets and moons.

3.8 Orbital Planes

The spin induced in a force-centre by its orbiting satellites, influences all the orbital planes in an orbital system.

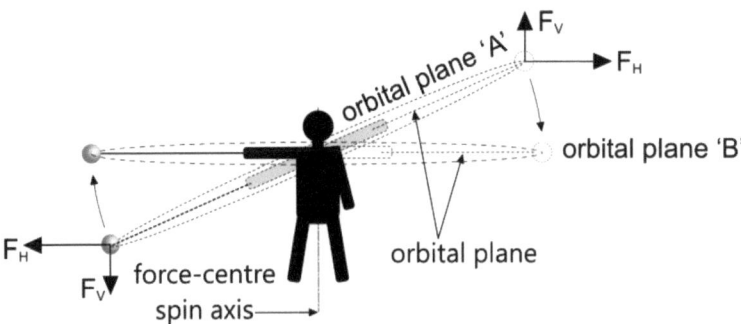

The rotational kinetic energy in a force-centre naturally causes the orbital planes of its satellites to settle at 90° to the force-centre's spin axis. This phenomenon can be demonstrated by attempting to swing a ball about *orbital plane 'A'* in which 'Fv' is non-zero, and subsequently through *orbital plane 'B'*, where 'Fv' is zero.

The same forces are at work in celestial orbital systems; any orbital plane will always settle at its lowest energy condition, which will always be at 90° to the force-centre's spin axis, unless an external force is preventing natural settlement.

So, once a dominant satellite has established its force-centre's spin axis, all the other satellites about the same force-centre will settle in the same plane.

3.9 How They Work

The mathematical laws for orbital systems were given to us by Copernicus, Kepler, Galileo and Isaac Newton before the end of the 17th century. They remain valid today, irrespective of the orbital system variables.

Refer to Appendix A-4 for a comprehensive analysis and verification of these laws.

3.9.1 Newton's Laws of Orbital Motion

These mathematical laws come in two parts:

    1) orbital shape and period

    2) satellite performance

The calculation results from one cannot be used to validate the other because the force-centre *alone* defines orbital shape and period. I.e. it is possible (in theory and in practice) to swap any satellite from its own orbit with that of another without altering the orbital periods or shapes. The only differences will be the satellite's performance.

This is why you can calculate the mass of a force-centre from the dimensional shape of any one of its satellite's orbits, but you cannot calculate the satellite mass in the same way.

A satellite's orbit (shape and period) is not altered by its own orbiting secondary satellites (e.g. moons).

All orbits are elliptical, the eccentricity of which must be; '$0 \leq e \leq 1$'

The following mathematical laws apply to all orbits, irrespective of eccentricity.

The key formula for orbital motion, originally defined by William Gilbert, and later described as follows by Isaac Newton:

    Potential Force: $F = G.m_1.m_2 / R^2$

which can be modified to define the following:

    Potential Energy: $PE = G.m_1.m_2 / R$

    Potential Acceleration: $g = G.m_1/R^2$

Alternative formulas for satellite velocity and potential acceleration:

    $v = h/R$

    $g = F/m_2$

Throughout any orbit, E is *always* constant, whilst PE and KE both vary, which is only possible because PE is negative and KE is positive:

    $E = KE - PE$

ORBITS

Constant of Proportionality:

$K = t^2/a^3 = (2.\pi)^2 / G.m_1$

Where:
G; Newton's *gravitational* constant
m_1; the *mass* of the force-centre
t; the total orbital period
a; half the length of the orbital major axis

in Linear orbits: $\frac{1}{2} . (2.\pi)^2 . t^2/d^3 = (2.\pi)^2 / G.m_1$

A useful tip from Kepler that was later verified by Newton is that the relationship between the swept area inside the ellipse for any given satellite's orbital period will always be identical.

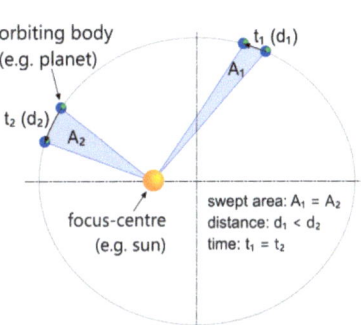

The calculations for all elliptical orbits are provided below (Chapter 3.9.2). They have been derived from Isaac Newton's *Philosophiæ Naturalis Principia Mathematica*, and they work perfectly for all elliptical orbits irrespective of satellite performance.

The Mathematical Variables:

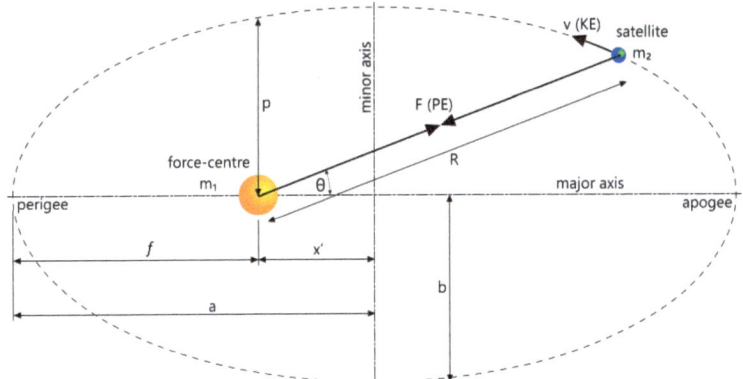

The variables in the above figure apply to the formulas provided in Chapter 3.9.2

ORBITS

3.9.2 Elliptical Orbital Formulas

Sym	Description	units
R^P	radius at orbital perigee ($R^P = 2.a - R^A$) [1]	kg
R^A	radius at orbital apogee ($R^A = 2.a - R^P$) [1]	kg
θ	angle through orbit [2]	c
t	orbital period	K

Table 3.9.2.1: *Input Data*

1) Once K has been established for the first orbit, only one of these values is required for all other satellites orbiting the same force-centre
2) $\theta = 0$ when satellite is at its apogee

Sym	Formula	Description	units
a	$(\check{R}+\hat{R})/2$ or $\sqrt[3]{t^2/K}$ [1]	major semi-axis	m
K	t^2/a^3 or $(2\pi)^2/G.m_1$	constant of proportionality [2]	s^2/m^3
f	$\check{R}$	Orbital Distance @ perigee	m
x'	a - f	orbit centre distance	m
e	$(-\check{R} + \sqrt{[\check{R}^2 - 4.a.\{\check{R}-a\}]})/2.a$	eccentricity	
b	$\sqrt{[a^2.(1-e^2)]}$	minor semi-axis	m
p	$a.(1-e^2)$	half-parameter	m
A	$\pi.a.b$	orbit total area	m^2
L	$\pi . \sqrt{[2.(a^2+b^2) - (a-b)^2 / 2.2]}$	orbital circumference	m
R	$p / [1 - e.\cos(\theta)]$	orbital radius @ θ	m
v	$2.A / t.R$	orbital velocity @ R	m/s
g	$-v.v^A / R.(1-e)$ [3]	gravitational acceleration	m/s^2

Table 3.9.2.2: *Orbital Shape*

1) This formula may be used to calculate a once K has been established for the first orbit
2) This value may be used to calculate $\check{R}$ or $\hat{R}$ for all other orbits around the same force-centre
3) v^A = satellite velocity at its orbital apogee

Sym	Formula	Description	units
m_1	$(2.\pi)^2 / G.K$	force-centre mass [1]	kg
m_2	input	satellite mass	kg

Table 3.9.2.3: *Body Mass*

1) Only one of these values is required for all orbits around the same force-centre once K has been established for the first orbit
2) Only required for a satellite between its orbital apogee and perigee

Sym	Formula	Description	units
F	$g.m_2$	potential force on satellite	N
PE	$F.R$	potential energy on satellite	J
KE	$\frac{1}{2}.m_2.v^2$	kinetic energy in satellite [1]	J
E	PE+KE	total energy [2]	J
p	$m_2.v$	satellite momentum	kg.m/s
h	$v.R$	Newton's motion constant	m^2/s
P	$F.v$	satellite power	J/s

Table 3.9.2.4: *Satellite Performance*

1) Does not include angular momentum, which plays no part on Newton's laws of orbital motion
2) This value does not change throughout a satellite's orbit (it is a constant for the satellite in its orbit)

ORBITS

The satellite mass (m_2) can be established if you know the potential acceleration (g_s) at a specified radius (r_s); e.g. at its surface:

either by using Isaac Newton's gravitational constant; $m_2 = -g_s \cdot r_s^2 / G$

or from the following formula if you've forgotten; G; $m_2 = m_1 \cdot g_s/g^P \cdot (r_s/R^P)^2$

ORBITS

3.9.3 Circular Orbital Formulas

The following Tables provide the mathematical formulas for the circular orbits of the proton-electron pair based upon their electrical charge attraction, according to Isaac Newton and Charles Augustine de Coulomb.

Sym	Description	units
m_e	electron mass	kg
m_p	proton mass	kg
T	temperature	K

Table 3.9.3.1A: *Input Data*

Sym	Formula	Description	units
R	X_R/T	orbital radius	m
e	0	orbital eccentricity	
A	$\pi.R^2$	orbital swept area	m^2
L	$2.\pi.R$	orbital path length	m
K	t^2/R^3	orbital constant of proportionality	s^2/m^3

Table 3.9.3.2A: Magnetic Orbital Shape

Sym	Formula	Description	units
v	$\sqrt{[T/X]}$	electron velocity	m/s
t	$2.\pi.R / v$	orbital period	s
a	$-v^2/R$	satellite centrifugal acceleration	m/s^2
F	$-k.(e/R)^2$	proton-electron pair potential force	J
PE	$F.R$	proton-electron pair potential energy	J
KE	$-PE/2$	satellite kinetic energy	N
F_c	$m_e.a$	satellite centrifugal force	N
E	$PE + KE$	total energy	J
h	$R.v$	constant of motion	m^2/s
P	$F.v$	electron power	J/s

Table 3.9.3.3A: Electric Orbital Performance

ORBITS

The following Tables provide the mathematical formulas for the circular orbits of the proton-electron pair based upon their magnetic charge (mass) attraction, according to Isaac Newton.

Sym	Description	units
m_e	electron mass	kg
m_p	proton mass	kg
T	temperature	K

Table 3.9.3.1B: *Input Data*

Sym	Formula	Description	units
R	X_R/T	orbital radius	m
e	0	orbital eccentricity	
A	$\pi.R^2$	orbital swept area	m^2
L	$2.\pi.R$	orbital path length	m
K	$\varphi.(2.\pi)^2 / G.m_p$	orbital constant of proportionality	s^2/m^3

Table 3.9.3.2B: Magnetic Orbital Shape

Sym	Formula	Description	units
a	$-G.m_p/R^2$	satellite centrifugal acceleration	m/s^2
v	$\sqrt{[a.R]}$	electron velocity	m/s
t	$2.\pi.R / v$	orbital period	s
F	$-G.m_e.m_p/R^2$	proton-electron pair potential force	J
PE	F.R	proton-electron pair potential energy	J
KE	-PE/2	satellite kinetic energy	N
F_c	$m_e.a$	satellite centrifugal force	N
E	PE + KE	total energy	J
h	R.v	constant of motion	m^2/s
P	F.v	satellite power	J/s

Table 3.9.3.3B: Magnetic Orbital Performance

The ratio between the two calculation methods; (electrical:magnetic) is the coupling ratio;

Difference	Symbol	Ratio	
a:a	φ	4.4074211179233E-40	$.../s^2$
v:v	$\sqrt{\varphi}$	2.0993858906650E-20	$.../s$
t:t	$\sqrt{\varphi}$	2.0993858906650E-20	$.../s$
KE:KE	φ	4.4074211179233E-40	$.../s^2$
PE:PE	φ	4.4074211179233E-40	$.../s^2$
F:F	φ	4.4074211179233E-40	$.../s^2$
$F_c:F_c$	φ	4.4074211179233E-40	$.../s^2$
E:E	φ	4.4074211179233E-40	$.../s^2$
h:h	$\sqrt{\varphi}$	2.0993858906650E-20	$.../s$

Table 3.9.3.4: Magnetic:Electrical Performance Ratios

ORBITS

3.9.4 Linear Orbital Formulas

Sym	Description	units
m^u	ultimate body mass; $m^u = k.e^2 / G.m_n.\varphi + m_n \geq 4.68687882273807E+48$ kg [1]	kg
m_1	force-centre mass [2]	kg
m_2	satellite mass [3]	kg
E_n	$1.63785606465701E-13$ [4]	J
v	galactic force-centre velocity (time-now)	m/s

Table 3.9.4.1: *Input Data*

1) universal mass > m^u; iterate to generate R_o
2) e.g. great attractor
3) e.g. galactic force-centre
4) physical constant

Sym	Formula	Description	units
N_n	$N_p.(1-1/(1+\psi))$	number of neutrons in m^u [1]	
E_t	$N_n.E_n$	total neutron energy in m^u	J
E	3% of E_t	ejection energy [2]	J
E_s	$E.m_2 / m^u$	satellite (ejection) kinetic energy	J
u	$\sqrt{[2.E / (m^u - m_1)]}$	ejection velocity (all ejected matter)	m/s
K	$(2\pi)^2 / G.m_1$ or $(\pi.t_o)^2 / \tfrac{1}{2}.R_o^3$	Newton's constant of proportionality	s^2/m^3
R_o	$\tfrac{1}{2}.u.t_o$	satellite terminal outward distance [3]	m
a	$-\tfrac{1}{2}.u^2/R_o$	outward satellite acceleration	m/s^2
g_o	a	gravitational acceleration @ R_o	m/s^2
t_o	$(4\pi)^2 / u^3.K$	time to R_o [4]	s

Table 3.9.4.2: *Performance @ Terminal Velocity = 0*

1) total number of neutrons in m^u (ψ = neutronic ratio of iron)
2) 3% represents the neutron loss in the 'little-boy' atom bomb
3) after which the satellite will return to m_{fc}
4) universal period = $2.t_o$

Sym	Formula	Description	units
t	$(v - u)/a$	universal age	s
R	$\tfrac{1}{2}.(v^2 - u^2)/a$	distance travelled from 'Big-Bang'	m
g	$-G.m_1 / R^2$	gravitational acceleration	m/s^2
KE	$\tfrac{1}{2}.m_2.v^2$	satellite kinetic energy	J
PE	$G.m_1.m_2 / R$	satellite potential energy	J

Table 3.9.4.3: *Satellite Performance @ Time-Now*

ORBITS

3.9.5 Centrifugal Force

Circular Orbits:

If you swing a ball - tied to a length of string - around your head, *centrifugal* force is pushing the ball away from you. But it also induces a tensile force in the string, pulling the ball towards you (*centripetal* force), *potential force*; the equivalent of *gravity*.

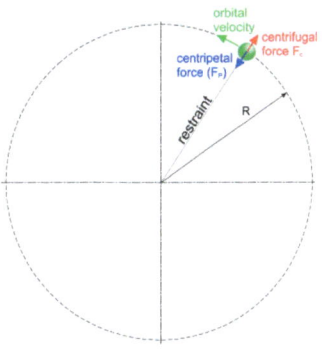

Christiaan Huygens gave us the mathematical relationship between this and its velocity in a circular orbit;

$$a = v^2/R \qquad \& \qquad F = m_2 \cdot v^2/R$$

where 'v' is its curvilinear velocity and 'R' is its orbital radius

Elliptical Orbits:

However, for elliptical orbits, the above orbital velocity (v) must be modified (v_c) to provide the correct gravitational acceleration, that would otherwise look like this:

$$a = v \cdot v^A / R \cdot (1-e)$$

where 'v^A' is the orbital velocity at the orbital apogee, 'e' is the orbital eccentricity and R is the orbital radius (at any orbital angle 'θ').

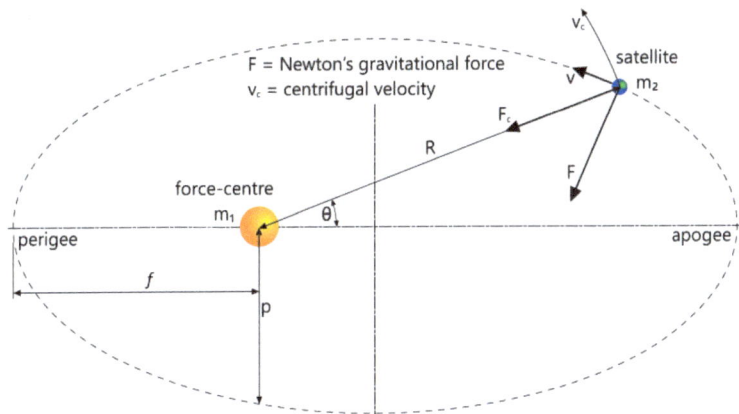

Centrifugal force in elliptical orbits may be calculated thus:

$$\alpha = \sqrt{[{}^4/_3 \cdot \pi]}$$
$$v_c = \kappa \cdot v$$
where: $\kappa = \sqrt{[\,(f \cdot \sin(\theta/2)^\alpha + p \cdot \cos(\theta/2)^\alpha) / (f \cdot \cos(\theta/2)^\alpha + p \cdot \sin(\theta/2)^\alpha)\,]}$
$$F_c = m_2 \cdot v_c^2 / R$$

which may be simplified at the orbital extremes as follows:

@ the perigee of an ellipse; $F_c = F \cdot f/p = F / (1+e)$
@ the apogee of an ellipse; $F_c = F \cdot p/f = F \cdot (1+e)$

ORBITS

3.9.6 Station-Keeping

This is a graphical representation of the restoration force on the earth 45° through its orbital path from its apogee.

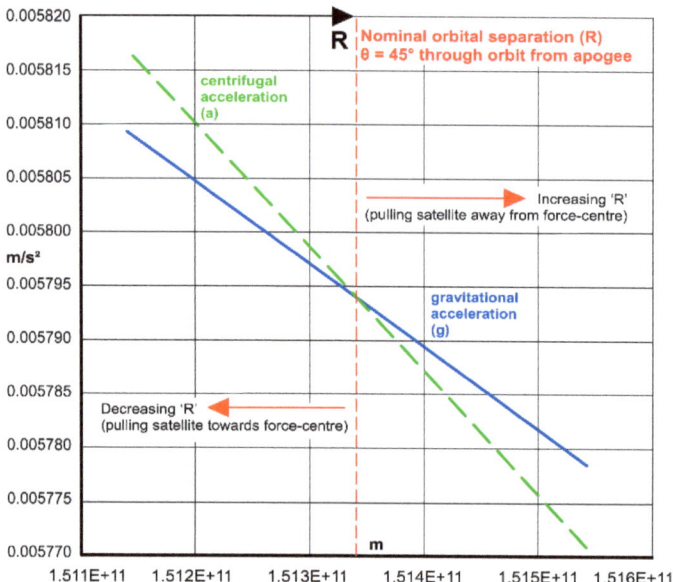

As the earth is pulled away from the sun, gravitational acceleration ($g = G.m_1/R^2$) increases faster than centrifugal acceleration ($a = v_c^2/R$), pulling the earth back towards the orbital path when the displacement force is released.

As the earth is pulled towards the sun, centrifugal acceleration (a) increases faster than gravitational acceleration (g), pulling the earth back towards the orbital path when the displacement force is released.

As you can see, exact balance occurs at the orbital path: at nominal orbital separation 'R'. And the following plot verifies the station-keeping and centrifugal force relationship.

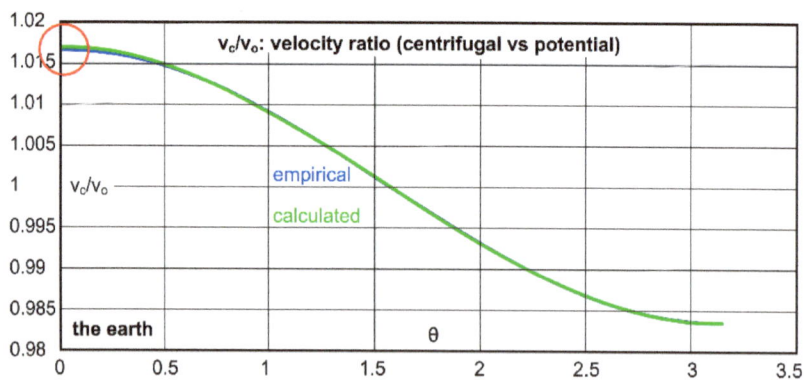

3.9.7 Orbital Precession

It is currently believed that satellites are temporarily influenced by other celestial bodies. The potential energy between the satellite and the external body will pull them together, out of their respective orbits temporarily altering their orbital paths. The satellite will accelerate as it travels towards the external body and decelerate (relative to the external influence) after passing it. However, these variations are effectively cancelled out as a result of Kepler's and Newton's 'equal time swept-area' law, station-keeping and the conservation of energy.

This is, however, incorrect. What actually happens is …

… whilst the orbital period is maintained, these temporary influences apply a torque; ($T_p = F_p \times R$) to the orbital axes causing a gradual rotation (of the orbit) about its force-centre. The frequency and magnitude of such influences define the rate of orbital precession.

It is this orbital rotation that is occasionally responsible for the impact between adjacent satellites and the creation of comets.

3.9.8 Orbital Planes

Natural orbits obey a fundamental law in that they *always* settle at their lowest energy condition.

All orbital planes about a single force-centre coincide because this is the lowest energy condition in each orbit:

$$F_h < \sqrt{[F_h^2 + F_v^2]}$$

F_h represents the potential force in orbit @ 90° to the force-centre's spin axis.

F_v represents an out-of-plane force when ≠ 90° to the force-centre's spin axis.

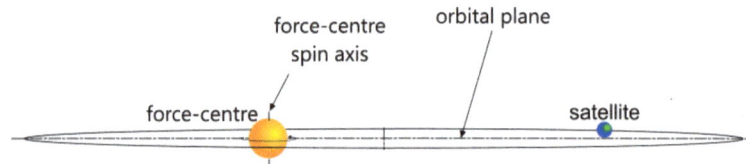

The satellite with the greatest kinetic and potential energy will dominate the force-centre's spin axis, after which, all lesser orbits about the same force-centre will eventually settle into the same plane; 90° to the force-centre's spin axis

Out-of-plane orbits are either newly trapped satellites that have yet to settle, or are subject to external forces preventing natural settlement.

4   Universal Orbits

Our universe is a linear orbital system; all of its satellites are said (NASA) to be currently travelling away from the 'Big-Bang' at about 230,000 m/s, at the periphery of an ellipsoid ≈4.353E+23 metres from its force-centre (the great attractor).

We know universal expansion is slowing down because all of its matter was initially ejected during the last 'Big-Bang' at ≈1.7735E+06 m/s; 0.592% of the velocity of EME. Because the current velocity of universal matter is [said to be] 0.077% of EME velocity, the heat (EME) in outer space cannot possibly be left over from the last 'Big-Bang', it can only be that generated and radiated from all of its active stars and planets. Heat radiated at the time of the last 'Big-Bang' must be 1.305E+32 metres away by now.

All the ejected matter in the universe is ≈4.676445E+48 kg. If our Milky Way is ≈1.7697E+41 kg - some will be larger and others smaller - there are ≈2.6425E+07 equivalent galaxies in our universe.

Because our galaxy is currently ≈4.353E+23 metres from its force-centre, the furthest distance of the universal ellipsoid must be ≈8.71E+23 metres away, and because EME travels at velocity 'c', the time for light to reach us from this distance (t = d/c) is ≈2.9E+15 seconds, which is only ≈0.668% of the age of our universe (4.34548E+17 seconds). Therefore, with a sufficiently high-definition camera, we should be able to see the entire universe; especially given that the light from the furthest reaches is from a much younger universe; closer to its force-centre and therefore it was radiated whilst closer to us.

Universal expansion will eventually cease (due to its force-centre's gravitational attraction) ≈4.993E+17 seconds (≈15.8219 billion years) from the 'Big-Bang', having reached a distance of ≈4.42755E+23 metres, after which it will travel back to re-accrete with the great attractor.

ORBITS

4.1 The Great Attractor

The great attractor is the residual (rubble) left over from the last 'Big-Bang', which will have quickly re-accreted into a single mass. It sits alone in empty space at the centre of the universe. Its mass is sufficient to slow down the outward travel of all ejected mass (galaxies) to their current velocity ($\approx$230000 m/s).

Because it comprises the same matter as the rest of the universe, it will consist of similar matter; predominantly iron, but its density will be considerably greater due to core pressure, say that of the heaviest stable element; lead.

Its properties may be summarised as follows:
    average density $\approx$ 11000   kg/m^3 (estimated)
    mass = 1.04336261407224E+46 kg
    surface radius = 6.09515366130185E+13 m
    surface area = 4.66851954870551E+28 m^2
    surface gravitational acceleration = 187,424,182.83575 m/s^2 (19,111,984 g)
    constant of proportionality = 5.66976153229753E-35 s^2/m^3
    potential acceleration:
        today = -3.67452423363053E-12 m/s^2
        @ v=0 $\approx$ -3.55196174079941E-12 m/s^2

It is expected that the temperature of the matter comprising the great attractor at the start of a universal period would be approximately 100K, falling to about 40K at its termination.

Due to its mass and low temperature, the great attractor is a significant source of universal fusion in much of its mass. It is also the reason we cannot see it in the night sky (dark).

4.2   The Milky Way

The mass of our sun (1.9885E+30 kg) is approximately 745.5 times greater than the sum of all its planetary mass (2.6676E+27 kg). Therefore, our stellar force-centre constitutes 99.866% of the mass of our solar system

If all the other solar systems orbiting Hades are similar to ours, and the stellar:galaxy mass ratio is similar to our solar system there are about 1.19E+08 - almost 120 million – equivalent solar systems in our milky way.

Its properties may be summarised as follows:
   total mass of ≈1.76808892E+41 kg
   comprises approximately 120 million [sun-equivalent] stars (most being dark).

Our solar system is in the outer reaches of the milky way, approximately 2.54E+20 metres from Hades.

ORBITS

4.3 Hades

Because we appear to have values for the sun's galactic orbit (NASA), we may calculate the properties of our own galactic force-centre, which I have named 'Hades' here in this publication to save me having to refer repeatedly to '*milky-way galactic force-centre*'.

Its properties may be summarised as follows:
>orbital eccentricity = 1
>orbital distance = 4.35308265895621E+23 m
>kinetic acceleration = -3.55196174079941E-12 m/s²
>average density ≈ 9000 kg/m³ (est.)
>mass = 1.76572018982E+41 kg
>surface radius = 1.673132E+12 m
>surface area = 3.517794E+25 m²
>surface gravitational acceleration = 4.209408E+06 m/s² (4.3E+05 g)
>spin energy = -8.75632E+48 J
>polar moment of inertia 1.97716E+65 kg.m²
>spin-rate = -9.41141E-09 c/s
>temperature ≤70K
>constant of proportionality = 3.35025744599744E-30 s²/m³

Initial:
>velocity = 1.7734984104E+06 m/s
>kinetic energy = 2.77067520630515E+53 J
>potential energy = 0 J

Today:
>orbital distance = 4.35308265895621E+23 m
>orbital period = 4.34548152E+17 s
>velocity ≈230000 m/s
>kinetic energy ≈ 4.67032990208312E+51 J
>potential energy ≈ 2.82435909300323E+53 J

@ v=0:
>orbital distance = 4.42754855113119E+23 m
>orbital period = 4.99301101717368E+17 s
>velocity = 0 m/s
>kinetic energy = 0 J
>potential energy = -2.77067520630515E+53 J

Hades is a linear-satellite; it cannot generate very much heat energy (through internal friction), because its force-centre does not rotate. That which it does generate will therefore be low, it will be cold (dark). The relative spin and potential energies of Hades and our sun when compared to their relative masses show us that heat generated and held by Hades must be negligible:

>Frictional heat; E_F = SE/PE / m   J/kg

ORBITS

Hades; $E_F = 8.75632E+48 \div 2.82436E+53 \div 1.76572E+41 = 18.40523$ J/kg

Sun; $E_F = 1.60098E+35 \div 9.5044E+40 \div 1.9885E+30 = 8.87966E+10$ J/kg

Therefore, our sun is generating 4.82453E+09 times more frictional energy than Hades
And because we know that our sun is generating fissionable energy in its core atoms, its core temperature must be; 623316124.717178 K
So, Hades' core temperature must be; 2.7255 K + 0.1292 K = 2.8547 K

And its EME radiation:
    $\lambda \approx 5.712591E-02$ m
    $A \approx 6.152892E-07$ m
    $f \approx 5.247925E+09$ /s
    $E \approx 1.874791E-22$ J

And given the sun's orbital radius ($\approx 2.5E+20$ m) and the size of an iron atom at 30°C ($\approx 7.5E-08$ m), searching for Hades in the night sky would be like looking for a black atom in the centre of a ten-metre diameter black iron disk. We know it exists because it is a fundamental law of nature that every orbital system *must* have a force-centre, and because it is cold it must be dark. That's why we can't see it; but it doesn't mean it isn't there!

4.4 Our Solar System

The mass of our sun (1.9885E+30 kg) is approximately 745.5 times greater than the sum of all its planetary mass (2.6676E+27 kg). Therefore, our stellar force-centre constitutes 99.866% of the mass of our solar system

Our solar system is in the outer reaches of the milky way, approximately 2.54E+20 metres from Hades.

Its properties may be summarised as follows:
        total mass ≈1.99116759E+30 kg
        comprises ≈20 planets and major comets

4.5 Our Sun

Our sun is a typical bright star. It has sufficient body and satellite mass to generate the internal frictional heat necessary for fission to occur in its core atoms.

Like all stars, it was originally a cold, dark, lone galactic satellite comprising similar matter to all other celestial bodies; it had no sub-satellites of its own. But as time (< 13.5 billion years) passed, it collected more and more of its own satellites from galactic comets and began to heat up. Eventually, it collected sufficient satellite mass to generate fissionable energy in its core atoms. Because spin energy only accounts for 3.43435%% of the sun's total energy, no matter how many more comets it traps, its heat (energy) is unlikely to increase significantly in the future.

Its properties may be summarised as follows:
  orbital eccentricity = 0.0159417437512301
  mass = 1.9885E+30 kg
  density (viscous matter) ≈ 6000 kg/m^3
  viscous diameter = 1.39E+09 m
  perigee distance = 2.4653729E+20 m
  apogee distance = 2.54525098196E+20 m
  orbital period = 7.258248E+15 s
  constant of motion = 5.43270958E+25 m^2/s
  constant of proportionality = 2.97491436434710E-19 s^2/m^3
  polar moment of inertia: 3.90008E+46 kg.m^2
  spin-rate: ω = 2.86533E-06 ^{c}/s
  core temperature = 623316124.717178 K
  surface temperature = 5788 K
  spin energy ≈ 1.60097718938751E+35 J
  fissionable energy ≈ 4.50156641838639E+36 J
  total energy ≈ 4.66166413732514E+36 J
  EME radiation ≈ 9.9252916097005E+33 J

It is currently claimed that our sun is creating elements from hydrogen (H$^+$) through fusion and that it is [apparently] growing in size with age.

The problem with this scenario is that fusion *increases density* and therefore *reduces size*. Moreover, why is Hades cold if fusion generates heat, as Hades is far more likely to generate fusion than our sun. Moreover, fusion only releases energy once; when it occurs, and as our sun's mass will not increase, it cannot have released fusion energy since the start of its life as a celestial body; almost 14 billion years ago.

Fissionable decay generates most of the sun's radiated heat energy and is therefore responsible for its hot hydrogen-helium atmosphere. The only reason we can see its atmosphere is because the hydrogen and helium at its surface comprises proton-electron pairs, which *are* capable of emitting electro-magnetic radiation. Our sun's atmosphere would be invisible (and cold) if its surface was lone protons (H$^+$).

4.6 Our Solar System

Sym	Mercury	Venus	Earth	Mars	Units
t	7600521.6	19413907.2	31558118.4	59354294.4	s
R^P	4.600120E+10	1.074770E+11	1.470950E+11	2.066550E+11	m
R^A	6.981450E+10	1.089346E+11	1.520942E+11	2.492139E+11	m
Orbital Dimensions					
a	5.790785019E+10	1.082057842E+11	1.495945981E+11	2.279344353E+11	m
e	0.02056137492	0.006735168186	0.01670914665	0.09335770306	
b	5.667054609E+10	1.082033299E+11	1.495737135E+11	2.269389619E+11	m
p	5.545967920E+10	1.082008757E+11	1.495528319E+11	2.259478361E+11	m
f	4.60012E+10	1.07477E+11	1.47095E+11	2.066550000E+11	m
x'	1.190665019E+10	7.287841549E+08	2.499598078E+09	2.127943533E+10	m
A	1.030966877E+22	3.678247729E+22	7.029445371E+22	1.625058045E+23	m^2
L	3.599700958E+11	6.798692829E+11	9.398649712E+11	1.429028790E+12	m
K	2.974914364E-19	2.974914364E-19	2.974914364E-19	2.974914364E-19	s^2/m^3
v^P	5.897421240E+04	3.525676706E+04	3.028600879E+04	2.649725040E+04	m/s
g^P	-6.271146496E-02	-1.148825844E-02	-6.133232761E-03	-3.107373324E-03	m/s^2
v^A	3.885846816E+04	3.478502382E+04	2.929053557E+04	2.197224924E+04	m/s
g^A	-2.722663689E-02	-1.118288438E-02	-5.736671536E-03	-2.136686702E-03	m/s^2
Masses					
m_1	1.9885E+30	1.9885E+30	1.9885E+30	1.9885E+30	kg
m_2	3.301100E+23	4.867370E+24	5.964520E+24	6.417100E+23	kg
Satellite Performance					
F^P	-2.070168170E+22	-5.591760450E+22	-3.658178805E+22	-1.994032536E+21	N
PE^P	-9.523022001E+32	-6.009856379E+33	-5.380998113E+33	-4.120767937E+32	J
KE^P	5.740543129E+32	3.025166886E+33	2.735455000E+33	2.252736683E+32	J
F^A	-8.987785104E+21	-5.443123593E+22	-3.421649078E+22	-1.371133223E+21	N
PE^A	-6.274777265E+32	-5.929443189E+33	-5.204129660E+33	-3.417054178E+32	J
KE^A	2.492298393E+32	2.944753696E+33	2.558586547E+33	1.549022924E+32	J
E	-3.782478872E+32	-2.984689493E+33	-2.645543113E+33	-1.868031254E+32	J
h	2.712884540E+15	3.789291553E+15	4.454920463E+15	5.475789281E+15	m^2/s
Inner Planets					

Input Data; ^{P} = perigee; ^{A} = Apogee;

ORBITS

Sym	Jupiter	Saturn	Uranus	Neptune	Units
t	*374335689.6*	*929596608*	*2651218560*	*5200329600*	s
R^P	*7.405200E+11*	*1.352550E+12*	*2.741300E+12*	*4.444450E+12*	m
R^A	8.156104E+11	1.501105E+12	2.997691E+12	4.548298E+12	m
Orbital Dimensions					
a	7.780652166E+11	1.426827696E+12	2.869495390E+12	4.496373960E+12	m
e	0.04825458812	0.05205792952	0.04467523818	0.01154796298	
b	7.771588241E+11	1.424893012E+12	2.866630380E+12	4.496074142E+12	m
p	7.762534876E+11	1.422960953E+12	2.863768230E+12	4.495774344E+12	m
f	7.405200000E+11	1.352550000E+12	2.741300000E+12	4.444450000E+12	m
x'	3.754521656E+10	7.427769562E+10	1.281953900E+11	5.192396004E+10	m
A	1.899659027E+24	6.387099181E+24	2.584205837E+25	6.351053352E+25	m^2
L	4.885880874E+12	8.958945951E+12	1.802057180E+13	2.825060890E+13	m
K	2.974914364E-19	2.974914364E-19	2.974914364E-19	2.974914364E-19	s^2/m^3
v^P	1.370590205E+04	1.015981492E+04	7.111398264E+03	5.495748646E+03	m/s
g^P	-2.419979478E-04	-7.254017694E-05	-1.765924516E-05	-6.718142608E-06	m/s^2
v^A	1.244404703E+04	9.154359018E+03	6.503164433E+03	5.370268285E+03	m/s
g^A	-1.994893535E-04	-5.889289434E-05	-1.476765724E-05	-6.414864102E-06	m/s^2
Masses					
m_1	1.9885E+30	1.9885E+30	1.9885E+30	1.9885E+30	kg
m_2	*1.898190E+27*	*5.683400E+26*	*8.681300E+25*	*1.024130E+26*	kg
Satellite Performance					
F^P	4.593580846E+23	-4.122748416E+22	-1.533052050E+21	-6.880251390E+20	N
PE^P	-3.401638488E+35	-5.576223370E+34	-4.202555585E+33	-3.057893329E+33	J
KE^P	1.782891576E+35	2.933255007E+34	2.195152879E+33	1.546602884E+33	J
F^A	-3.786686960E+23	-3.347118757E+22	-1.282024628E+21	-6.569654772E+20	N
PE^A	-3.088461391E+35	-5.024378011E+34	-3.843113407E+33	-2.988074714E+33	J
KE^A	1.469714479E+35	2.381409647E+34	1.835710700E+33	1.476784269E+33	J
E	-1.618746912E+35	-2.642968364E+34	-2.007402707E+33	-1.511290445E+33	J
h	1.014949459E+16	1.374165767E+16	1.949447606E+16	2.442558007E+16	m^2/s
Outer Planets					
Input Data; ^{P} = perigee; ^{A} = Apogee;					

ORBITS

Sym	Mercury	Eris	Haumea	MakeMake	Units
t	7824384000	17610912000	8966073600	9754300800	s
R^P	4.436820E+12	5.6714E+12	5.22874E+12	5.77298E+12	m
R^A	7.371060E+12	1.46082E+13	7.7015E+12	7.90439E+12	m
Orbital Dimensions					
a	5.903940173E+12	1.01398E+13	6.46512E+12	6.83869E+12	m
e	0.02484984824	0.440679758	0.191238746	0.155834797	
b	5.718747063E+12	9.10214E+12	6.3458E+12	6.75514E+12	m
p	5.539363037E+12	8.17067E+12	6.22868E+12	6.67261E+12	m
f	4.436820000E+12	5.6714E+12	5.22874E+12	5.77298E+12	m
x'	1.467120173E+12	4.46841E+12	1.23638E+12	1.06571E+12	m
A	1.060700342E+26	2.8995E+26	1.28888E+26	1.4513E+26	m^2
L	3.651627701E+13	6.04983E+13	4.02476E+13	4.27067E+13	m
K	2.974914364E-19	2.97491E-19	2.97491E-19	2.97491E-19	s^2/m^3
v^P	6.110837548E+03	5806.056018	5498.490967	5154.546835	m/s
g^P	-6.741268858E-06	-4.12577E-06	-4.8539E-06	-3.98185E-06	m/s^2
v^A	3.678261332E+03	2254.105842	3733.060617	3764.628897	m/s
g^A	-2.442448046E-06	-6.21857E-07	-2.23735E-06	-2.12397E-06	m/s^2
Masses					
m_1	1.9885E+30	1.9885E+30	1.9885E+30	1.9885E+30	kg
m_2	1.303E+22	1.66E+22	4.006E+21	4.4E+21	kg
Satellite Performance					
F^P	-8.783873322E+16	-6.84877E+16	-1.94447E+16	-1.75202E+16	N
PE^P	-3.897246483E+29	-3.88421E+29	-1.01671E+29	-1.01143E+29	J
KE^P	2.432853160E+29	2.79795E+29	6.05575E+28	5.84526E+28	J
F^A	-3.182509804E+16	-1.03228E+16	-8.96283E+15	-9.34547E+15	N
PE^A	-2.345847182E+29	-1.50798E+29	-6.90273E+28	-7.38703E+28	J
KE^A	8.814538586E+28	4.21722E+28	2.79133E+28	3.11793E+28	J
E	-1.464393323E+29	-1.08626E+29	-4.1114E+28	-4.26909E+28	J
h	2.711268625E+16	3.29285E+16	2.87502E+16	2.97571E+16	m^2/s
Peripheral Planet and Comets					
Input Data; ^{P} = perigee; ^{A} = Apogee;					

A-1 Glossary

Apogee (aphelion)	the point at which a satellite's orbit passes furthest from its force-centre. The term apogee normally applies to lunar orbits, and aphelion is normally used for planetary orbits, but they mean the same thing.
atom	a collection of deuterium and tritium atoms that were fused together in core of massive, cold bodies.
atomic particles	the proton and the electron
Big-Bang	The eruption of matter from the ultimate body due to neutron-neutron interaction in its core atoms; a chain reaction
charge (electrical)	the electrical potential of an atomic particle
charge (magnetic)	the magnetic potential of an atomic particle (what we currently refer to as mass)
EME	the electro-magnetic energy generated by a proton-electron pair
energy	the distance over which a force is applied
force	the effort required to change the velocity and/or direction of a mass
force-center	the central body (anchor) of an orbital system.
galactic force-centre	the force-centre of a galaxy.
gravity	non-polar magnetic attraction
(the) great attractor	the residue left over from the last 'Big-Bang', that is slowing down the outward travel of all universal galaxies, and will eventually cause them to reaccrete into another ultimate body.
Hades	the name adopted here for the milky way's force-centre
heat	the electro-magnetic energy absorbed by a body's atoms
hydrogen	a proton-electron pair
kinetic energy	the energy in a mass moving at a constant velocity
mass	[non-polar] magnetic charge

massive body	cold bodies with sufficient mass to generate the core pressure necessary to fuse atoms; e.g. galactic force-centres, the great attractor and the ultimate body
matter	collection of atoms
neutron	proton-electron pair united through high temperature
neutronic	the condition of a proton-electron pair at the instant of its union as a neutron
orbit	the curvilinear path followed by a satellite about its force-centre
orbital system	a force-centre and all of its satellites
Perigee (perihelion)	the point at which a satellite's orbit passes closest to its force-centre. The terms perigee normally applies to lunar orbits, and perihelion is normally used for planetary orbits, but they mean the same thing.
potential energy	the straight-line attraction or repulsion between masses
proton-electron pair	a proton with a single orbiting electron
satellite	an orbiting body
(the) ultimate body	all universal matter that has reaccreted at the end of a universal period due to the gravitational attraction of the great attractor. It constitutes all the matter in the universe.
universal period	the period of time between 'Big-Bangs' (64.75 bn years)

THE MATHEMATICAL LAWS OF NATURAL SCIENCE

A-1.1 Terminology

An **orbit** is the path followed by a satellite around its force-centre. For example, our moon is in orbit around our Earth (a planet), which is in orbit around our sun (a star), which is in orbit around Hades (a galactic force-centre). Electrons orbit their protons (a proton-electron pair).

An orbiting body (or *mass*) is referred to as a **satellite** and the body about which it orbits is referred to as its **force-centre**.

Solar Orbit: A star's orbital path around its galactic force-centre.

Planetary Orbit: A planet's orbital path around its star.

Lunar Orbit: A moon's orbital path around its planet.

Atomic Orbit: An electron's orbital path around its proton.

An **orbital system** is a force-centre with all of its satellites and sub-satellites, all of which have group names such as:

Collectively, everything (including the force-centre) orbiting a …

… galactic force-centre is called a **galaxy**

… star is called a **solar system**

… planet is called a **lunar system**

… proton is called a **proton-electron pair**

An orbit is *always* a perfect ellipse, exactly as Johannes Kepler stated. An ellipse can be any two-dimensional (flat) elliptical shape including a circle.

There is a major difference between a genuine elliptical orbit, i.e. one in which its axes *are not* identical in length, and a circular orbit, i.e. one in which both its axes *are* identical in length:

Satellites following **elliptical orbits** (e.g. stars, planets, moons, comets, etc.) keep going because of the potential and kinetic energies (Newton's constant of motion) between a satellite and its force-centre.

Satellites following **circular orbits** (e.g. electrons, geo-political, etc.) keep going because they provide their own kinetic energy.

Velocity in orbits refers only to the curvilinear motion of a satellite in its path around its force-centre. It does not refer to rotational (angular) motion in a satellite or its force-centre.

Mass is magnetic charge.

Gravity is the potential energy generated by magnetic charge.

Force is energy per unit distance.

Energy is a force applied over a specified distance.

Kinetic energy is the energy in a satellite due to its velocity.

Potential energy is the attractive/repulsive energy between a satellite and its force-centre.

Planetary Spin is the angular velocity (radians per second) in a body rotating about an axis that passes through its centre of *mass*.

A-2 References

There is little in today's scientific literature that has, or can, help to resolve the mathematical laws of natural science. Therefore, apart from the achievements of the heroes listed in Appendix A-4 of this book, most of these laws have been established as a result of the work done in the previous publications listed below:

The Mathematical Laws of Natural Science; Keith Dixon-Roche; 9-798610-294490

The Universe; Keith Dixon-Roche; 978-1-70753-878-2

It is important to note here that most of the sources used in this work are from work done by pre-20th Century scientists that are universally known and available from sources too numerous to mention here.

THE MATHEMATICAL LAWS OF NATURAL SCIENCE

A-3   Useful Formulas

Equidistant arc-length between 'n' points on the surface of a sphere:

$d = \pi \cdot A / C \cdot n$

where C is the circumference of the sphere

Linear distance across arc-length 'd' (above):

$\ell = 2 \cdot R \cdot \sin(\tfrac{1}{2} \cdot d/R)$

but if you know 'ℓ' and need to find 'n':

$n = \pi / \mathrm{Asin}(\tfrac{1}{2} \cdot \ell/R)$

and if $\ell = R$:

$n = \pi / \mathrm{Asin}(\tfrac{1}{2}) = 6$

Lorentz's Equation (magnetic force or field strength):

$F = q \cdot v \cdot B$

Which becomes:

$F = q \cdot g \cdot R \cdot B$

for the laws of orbital motion

Where:

q is the total electrical charge = $q_1 \cdot q_2 / m_e \cdot (q_1 + q_2)$

v = relative velocity (electrical circuits)

g = gravitational attraction between m1 & m2

R = radial separation between m1 & m2

$B = \mu_o \cdot e / R_n = R_n \cdot m_e / e^2 \cdot e / R_n = m_e / e = 1/RC$ kg/C

RC is the relative atomic charge capacity of an electron {C/kg}

$B = 1/RC = 5.685634367312\mathrm{E}{-}12$ kg/C

THE MATHEMATICAL LAWS OF NATURAL SCIENCE

A-4 Newton's Orbital Laws

The four principal agents for the theories of planetary motion were Copernicus, Kepler, Galileo and Newton. Between them, they defined the behaviour of orbiting satellites, moons and planets that remain valid even today.

A-4.1 Nicolaus Copernicus (1473 to 1543)

Copernicus stated that; contrary to religious doctrine, the sun does not orbit the earth, but all the planets in the solar system orbit the sun. He was so concerned for his safety regarding this claim, however, that he arranged for the publication of his findings to be deferred until after his death.

A-4.2 Johannes Kepler (1571 to 1630)

Kepler used Tycho Brahe's (1546 to 1601) observational data to show that the planets not only orbited the sun, just as Copernicus had previously claimed, but that their orbital paths were ellipses. Kepler also stated that the time taken to traverse between any two points on this elliptical curve is proportional to the swept area:

i.e; $t_1/A_1 = t_2/A_2$

Whilst he did not provide a mathematical proof for his swept area theory, he understood it. It was later confirmed by Isaac Newton.

A-4.3 Galilei Galileo (1564 to 1642)

Galileo is best known for his physical evidence of celestial bodies (moons) orbiting other planets, revealed in his book; Dialogue Concerning the Two Chief World Systems (frequently referred to as the 'Dialogue'), therein declaring Copernicus correct and finally quashing over a thousand years of religious dogma that stated all celestial bodies orbit the earth. In return for his findings, he was put under permanent house arrest, but only after being threatened with death if he didn't recant this claim.

However, it was during his confinement that Galileo completed his most important work, his laws of motion, one of which states that a body fired from the surface of the earth would follow a parabolic curve back to its surface.

This claim may be demonstrated by comparing Galileo's mathematically correct parabola with a projectile trajectory calculation:

$x(t) = A.t + B$

$y(t) = C.t + D - \frac{1}{2}.g.t^2$

If B and D are zero {i.e. v occurs at t = 0}:

$x(t) = v.\cos(\alpha).t$

$y(t) = v.\sin(\alpha).t - \frac{1}{2}.g.t^2$

Where:

v = initial velocity

A = initial horizontal velocity {i.e.; A = v.Cos(α)}

B = offset horizontal distance from t = 0

C = initial vertical velocity {i.e.; C = v.Sin(α)}

D = offset vertical distance from t = 0

This plot shows the projectile trajectory (curve) superimposed on two alternative parabolic curves, one of which passes through the same latus rectum and the other being the best parallel match.

Whilst the parabolic path is not strictly correct, it is stunningly close, demonstrating that given the limited information and facilities available to Galileo at his time, he was a very capable mathematician.

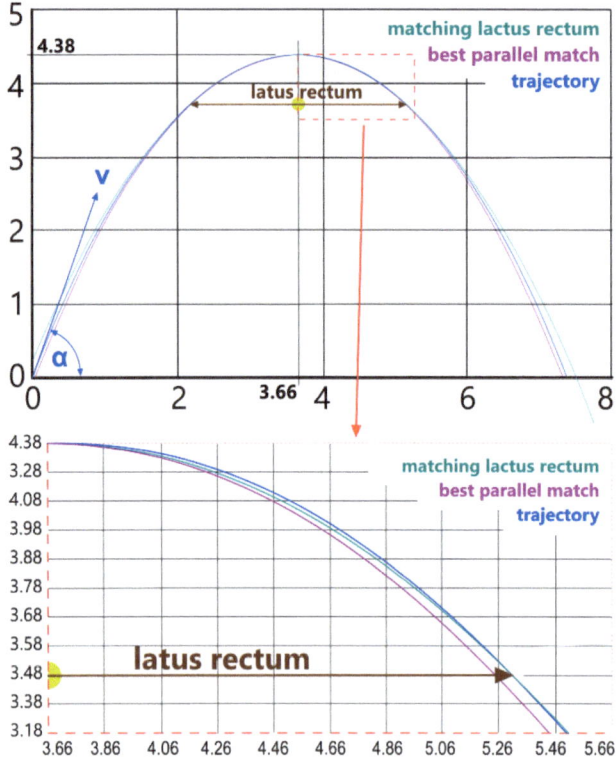

A-4.4 Isaac Newton (1642 to 1727)

Along with [his explanation for] *gravity*, Newton used [his creation of] calculus to mathematically prove the theories previously generated by Copernicus, Kepler and Galileo. In 1687 he published his results under the heading Philosophiæ Naturalis Principia Mathematica (first of three issues), probably the most important scientific work ever produced.

In his Principia, Newton discusses the alternative curves that describe the elliptical paths followed by an orbiting body. However, the parabolic and hyperbolic curves can only be responsible for paths followed by a body (e.g. a galactic comet) travelling towards a force centre from well outside its influence, sufficiently close to fall under its influence, pass around the force centre and then travel back out of its influence. A complete orbit, i.e. that of a satellite must be an ellipse.

$$F = \frac{G \cdot m_1 \cdot m_2}{r^2}$$

As a result of this work, Newton defined the fundamental relationship (G) between attracting bodies in which the potential force (F) is directly proportional to the inverse of the square of the distance (R) between the attracting bodies.

Whilst a value for 'G' was never established by Newton, despite it being of special importance to his theories, it has been estimated many times since the publication of Principia, varying between 6.67E-11 and 6.76E-11 N.m²/kg²

The minimum and maximum radial distances between the earth and sun (@ a & b respectively) are assumed to be as defined in the Earth-Sky fact sheet (https://earthsky.org/). Therefore, using Newton's theories and true value for 'G', the principal properties of the earth's orbit are as follows:

a = 1.495945981E+11m (R + R)/2)

b = 1.495737135E+11 {√[a².(1-e²)]}

e = 0.01670914665 {a.e² + R.e + R - a = 0}

p = 1.495528319E+11m {a.(1-e²)}

f = 1.47095E+11m {a.(1-e)}

x' = 2.499598078E+09m {a-f}

R = 1.47095E+11m to 1.520941962E+11m

F = 3.658178805E+22 to 3.421649078E+22N

v = 30286.008788376 to 29290.53557m/s

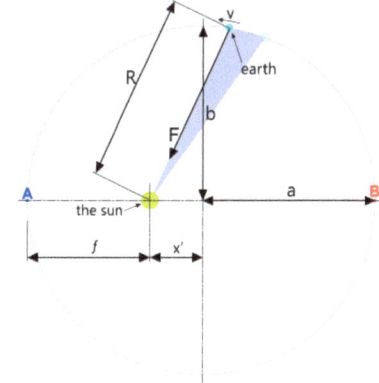

Newton's creation of Calculus allowed him to generate formulas for non-linear versions of Galileo's relationships for distance (s), time (t), velocity (v) and acceleration (a) as follows:

$s = ut + \tfrac{1}{2}at^2$

$\delta s/\delta t = v = u + at$

$\delta^2 s/\delta t^2 = \delta v/\delta t = a$

A-4.5 Proof (elliptical orbits)

By applying calculus, Newton was able to generate the non-linear formulas necessary to complete his theories concerning the elliptical (conic) path of orbiting bodies, which was proven as follows:

Assume an ellipse and the planet is passing the x-axis @ 'A' (y = 0)

x component = R {a}

y component = v/ω {b}

$x(t) = R.\sin(\omega.t)$

$y(t) = (v/\omega).\cos(\omega.t)$

From: $\sin^2(\omega.t) + \cos^2(\omega.t) = 1$

$y(t)^2 / (v/\omega)^2 = 1 - \sin^2(\omega.t)$

$\sin^2(\omega.t) = 1 - y(t)^2 / (v/\omega)^2$

$x(t)^2 / R^2 = \sin^2(\omega.t) = 1 - y(t)^2 / (v/\omega)^2$

$x(t)^2 / R^2 = 1 - y(t)^2 / (v/\omega)^2$

$x(t)^2 / R^2 + y(t)^2 / (v/\omega)^2 = 1$

An ellipse!

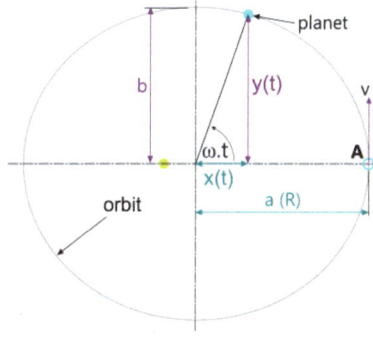

A-4.6 Euclidean Geometry (equal areas)

Whilst Kepler had already predicted the equal-swept-area-with equal-orbital-time theory, it had still not been mathematically proven by the time Newton was writing his Principia. Newton did this using Euclidian geometry.

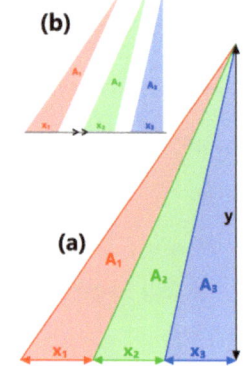

The areas of each triangle; A_1, A_2 & A_3 are all equal if the base widths; x_1, x_2 & x_3 are equal, which can be proven as follows:

Let y=6 and x_1, x_2 & x_3 all equal 3 (a)

The area of a triangle: $A = x.y/2$

$A_1 = 6 \times 3 \div 2 = 9$

$A_2 = 6 \times (3+3) \div 2 - A_1 = 9$

$A_3 = 6 \times (3+3+3) \div 2 - A_1 - A_2 = 9$

Therefore, all the areas are equal (i.e. 9)

The same applies to triangles with equal bases between parallel lines (b)

He then applied this to the conservation of energy

THE MATHEMATICAL LAWS OF NATURAL SCIENCE

A-4.6.1 Proof (conservation of energy & equal time-swept area)

Newton's proposition diagram for his proof of Kepler's 'equal-areas-equal-time' theory is shown here, where the following instructions describe its construction (author's words):

1) Divide time [of orbit] into equal parts [represented by equal swept areas {triangles}]

2) Assume the line A-B describes the linear path of the body if unconstrained by potential attraction

3) The same body would then continue to B-c

4) Assume that the body is attracted by a central-force (S) and diverted from its right line (B-c) in a direction parallel to V-B as far a C

5) Continue to generate similar triangles (S-A-B) following the points D, E, F, etc.

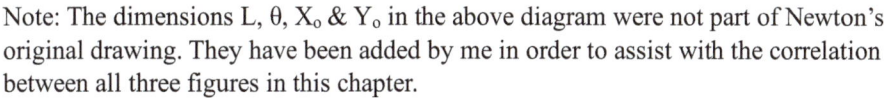

Note: The dimensions L, θ, X_o & Y_o in the above diagram were not part of Newton's original drawing. They have been added by me in order to assist with the correlation between all three figures in this chapter.

Newton was therefore stating that all swept areas (triangles SAB, SBC, SCD, SDE, SEF, etc.) must be equal.

The difficulty in generating the above diagram is knowing how far along the line c-C that C occurs in order to ensure that each subsequent area remains equal.

This can be achieved using the process described in the figure below, where the blue variables are entered (X_0, Y_0, L, θ), all the green variables (X_1, Y_1, R_1, r_2, A_2, ε_2, X_0, α_1) can be easily calculated using the blue variables, and the red variables (h, R_2, α_2, X_2, Y_2) may be determined using the formulas provided.

$h = r_2 . \tan(\varepsilon_2)$
$R_2 = 2.A/h$
$\alpha_2 = [h/R_2 . \sin(½\pi - \varepsilon_2)]$
$X_2 = R_2 . \cos(\Sigma\alpha)$
$Y_2 = R_2 . \sin(\Sigma\alpha)$

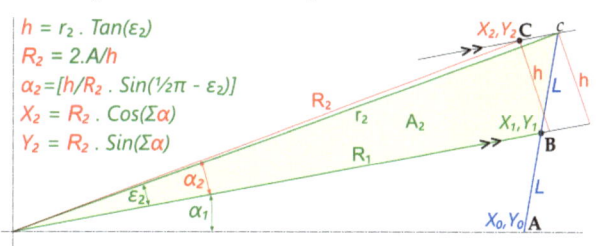

Newton claimed that if you reduce length L 'in infinitum' and join up the dots (X,Y co-ordinates) you produce a curved line thereby demonstrating that a centripetal force is continually acting on the body in the direction of the force centre and the triangular areas will always be proportional to the time passed by the body traversing each triangle; QED

Newton's constant of motion 'h' is not to be confused with the perpendicular distance 'h' shown in the above diagram; they are neither the same nor in any way connected.

55

Newton's diagram is less easily seen from his words and his fairly simple diagram than if you actually complete it, and repeat it for ever smaller values of L through to 360°:

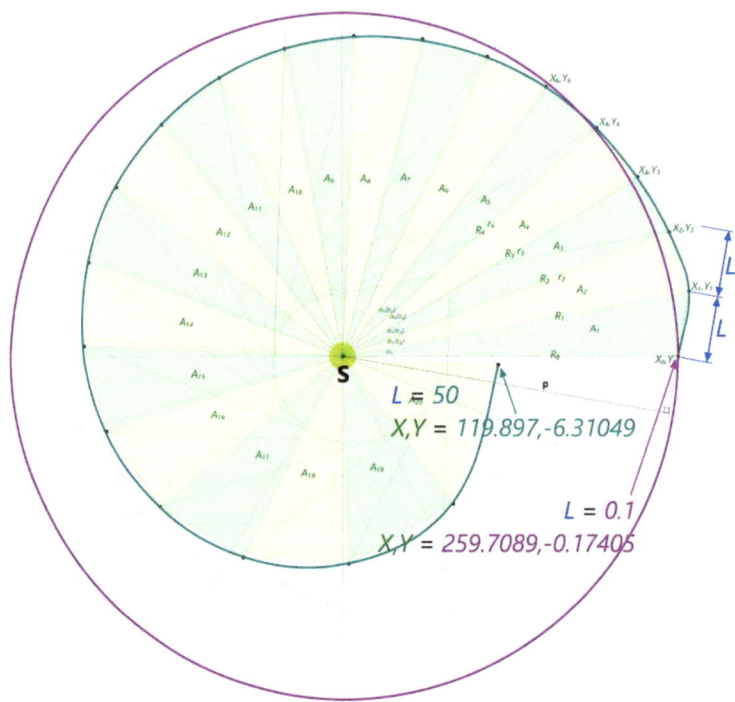

A calculation was carried out using the following input data:

$X_0, Y_0 = $ **260,0**

$L = 50$

$\theta = 100°$

As can be seen above, Newton's diagram does indeed produce a curve, exactly as he claimed, and following this through a sequence of diminishing values for L from 50 to 0.1, the following X,(Y) co-ordinates are achieved immediately prior to reaching 360° …

L	X	(Y)
50	119.74535317	(-5.41947224)
25	179.77612136	(-11.51597352)
10	229.5157957	(-9.468518876)
1	257.067565265	(-0.349111926)
0.1	**259.707825986**	**(-0.076479042)**

THE MATHEMATICAL LAWS OF NATURAL SCIENCE

Note: the 'Y' co-ordinate is in parenthesis because it is simply a resultant. The trend is demonstrated by the 'X' co-ordinate.

... from which it isn't difficult to anticipate where X (& Y) will end up if L is diminished in infinitum [i.e.: X(,Y) = **260(,0)**], making the final shape a circle and thereby proving that:

a) the path of the body is continuous (conservation of energy and momentum)

b) the orbital time passed by the body is proportional to the swept area (triangle)

c) Newton's calculus can be used to determine the properties of the path {'in infinitum'}

This result does not mean that the orbital path is circular, simply that it is continuous.

Corollary 1

Newton's first Corollary (to the above proof) states that the velocity of the body (v), represented by L, at positions A, B, C, D, E, F, etc. (Newton's diagram) is inversely proportional to the perpendicular distance of its tangent from the force centre (p)

Newton also stated that; v multiplied by p is a constant, i.e. his constant of motion (h), which is the angular momentum without the *mass* component.

Using the above 'in infinitum' argument it can be seen in the following table where these calculations have been carried out for successively reduced values of L between the start and end of the orbit

(h_o @ 0°, h < 360°), 'h' does indeed become a constant:

L	H_o	h
50	47.46494162	106.803277
25	24.20484775	35.53390597
10	9.782449005	11.14661354
1	0.984150369	0.996040786
0.1	**0.098474198**	**0.098591562**

A-4.6.2 Centripetal Force

Centrifugal acceleration (according to Christiaan Huygens {1629 to 1695}): $a = R.\omega^2$

where $\omega = 2.\pi/t$

$a = \sqrt{[(R.\omega^2)^2 + (R.\alpha)^2]}$

with constant angular momentum; $\alpha = 0$

$a = R.\omega^2$

$a = R.(2.\pi/t)^2$

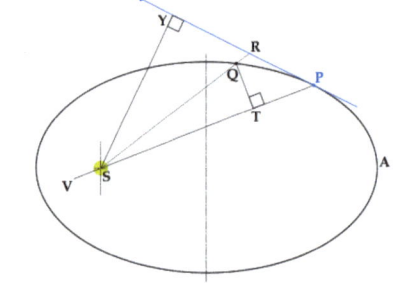

Centrifugal force:

$F = m.a$

$F = m.R.(2.\pi/t)^2 = 4.\pi^2.m\ (R/t^2)$

Through his inverse rules, Newton shows that the centripetal force (F) between the orbiting body and the force-centre (S);

$F = SP^2 . QT^2 / QR$

$PR = v^P . \delta t$

where; v^P is the velocity of the body at P and δt is the time taken for the body to travel from P to Q

$QR = (F^P / 2m).\delta t^2$

where; F^P is the centripetal force on the body at P

$F = QR . (2m / \delta t^2)$

where; F^P is the centripetal force on the body at P

$\delta t = PR/v^P = PR / (h/SY) = PR . SY / h$

where h is Newton's constant of motion (see Corollary 1 above)

Therefore, the centripetal force (F) can be calculated as follows:

$F = QR . (2.m / (PR . SY / h)^2)$
$\quad = QR . (2.m / (PR^2 . SY^2 / h^2))$
$\quad = QR . (2.m.h^2 / (PR^2 . SY^2))$
$\quad = QR.2.m.h^2 / PR^2.SY^2$

Newton preferred the calculation in geometric form by setting $2.m.h^2$ as a constant (k):

$F = k.QR / (PR.SY)^2$

A-4.6.3 Distance Between a Satellite & its Force-Centre (R)

The separation (distance) between an orbiting body and its force centre, can be found by using general elliptical equation:

$R = a.(1-e^2) / (1-e.\cos(\theta))$

Where; 'R' is the distance between the satellite and its force-centre at 'θ', 'θ' is the angle of the satellite's orbital progress from its apogee and 'e' is the orbital eccentricity

The force centre is not at the centre of an ellipse but at its focus (S)

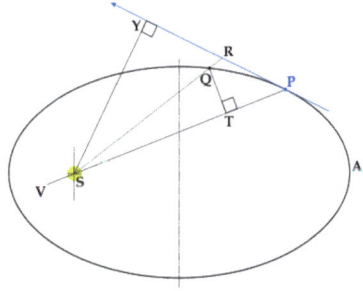

A-4.6.4 The Inverse Square Law

Proposition XI: "If a body revolves in an ellipse; it is required to find the law of the centripetal force tending to the focus of the ellipse"

Using similar geometric arguments as above (A 5.6.1 to 3) Newton worked out that the force between an orbiting body and its force centre is proportional to the inverse of their separation (the distance between them): $F \propto 1/R^2$ i.e. $F = K/R^2$

where:

the constant of proportionality: $K = G.m_1.m_2$ i.e. $F = G.m_1.m_2 / R^2$

where G is a constant and m_1 and m_2 are the masses of the force centre and the orbiting body

This same relationship ($F \propto 1/R^2$) also applies to parabolas and hyperbolas as well as the ellipse

The above constant of proportionality (K) can also be written as;

$K = m.h^2/p$

Where 'h' and 'p' are defined in Corollary 1 above and m is the mass of the orbiting body

i.e. $F = (m.h^2/p).(1/R^2)$

In the first formula, you can resolve the problem knowing the mass of the bodies

In the second formula, you can resolve it knowing the velocity and mass of the orbiting body and the parameter of its curve (p)

Both of the above F calculations produce the same result;

e.g. the following centripetal force occurs in the earth's orbit, 0.000175° from the major semi-axis:

$G = 6.67359232004332E-11$ (gravitational constant)

$m_1 = 1.9885E+30$ (sun mass)

$m_2 = 5.964519768E+24$ (earth mass)

$R = 1.5209420E+11$ (distance between mass centres)

$F = G.m_1.m_2 / R^2 =$ **3.421649078E+22** (centripetal)

$h = 4.454920463E+15$ (constant of motion - see Corollary 1 above)

$m = 5.964519768E+24$ (earth mass)

$p = 1.495528319E+11$ (ellipse parameter)

$R = 1.5209420E+11$ (distance between mass centres)

$F = m.h^2 / p.R^2 =$ **3.421649078E+22**

A-4.6.5 Orbital Period

Proposition XV: "The same things being supposed, I say, that the periodic times in ellipses are as the $3/2^{th}$ power of their greater axes"

This means that if the major semi-axis of an ellipse is 'a' (A 5.5) and the time taken for a body to orbit the elliptical path is 't' then the relationship between the two is:

$t \propto (2.a)^{1.5}$ or $t^2 \propto (2.a)^3$

Therefore; $t = K \cdot a^{1.5}$

Where K is the constant of proportionality, which is dependent on the properties of the force-centre.

This is actually Kepler's third law

A-4.6.6 Constant of Proportionality

To determine 'K' (the constant of proportionality for $t = K \cdot a^{1.5}$

...

$K = t^2 / a^3 \ \{s^2/m^3\}$

now we know ...

... that the earth travels around the sun in 31558149s

... the earth's semi-major orbital x-axis is 1.495945981E+11m

Therefore:

$t^2 / a^3 =$ **2.974914364E-19** $\{s^2/m^3\}$

$G = 6.67359232E\text{-}11 \ \{N.m^2/kg^2 = kg.m.m^2 / s^2.kg^2 = m^3 / s^2.kg\}$

$m_1 = 1.9885E\text{+}30$ kg (the *mass* of our sun)

$1 / m_1.G = 7.535546116E\text{-}21 \ \{s^2.kg / m^3/kg = s^2/m^3\}$

2.975944645E-19 $\div 7.538155846E\text{-}21 = 39.47841760436$

$\sqrt{39.47841760436} = 6.2831853071796 = 2.\pi$

Therefore:

$K = (2\pi)^2 / G.m_1 = (2\pi)^2 \div 6.67359232E\text{-}11 \div 1.9885E\text{+}30$

$\quad =$ **2.974914364E-19** s^2/m^3

i.e.;

$K = (2\pi)^2 / G.m^{fc}$

where m^{fc} is the *mass* of the force-centre

The above calculation, based upon NASA's data for the sun and the earth's orbit, gives an error margin of 0

A-4.6.7 Alternative Velocity Calculation

A much simpler orbital velocity calculation method is based upon Kepler's 'swept-area = time' rule

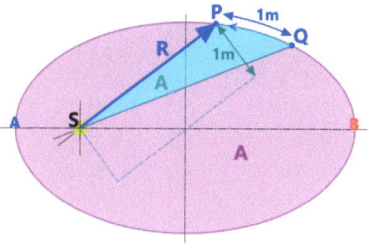

Using the earth's orbit as an example:

Earth's total orbital area (A) is 7.029445371E+22m² and it takes 31558149s (t) to complete

The swept area (A) is equal to ½.R x 1m

The velocity of the orbiting body at any given distance between the centres of *mass* (P & S) is calculated as follows:

v = 2.A / t.R {m² / s.m = m/s}

By way of verification:

The earth's maximum velocity occurs when R = 1.47095E+11 m (@ A)

v = (2 x 7.029445371E+22) ÷ (31558118.4 x 1.47095E+11)

   = **30286.008788376** m/s

30286.008788376 m/s (calculated using; h=v.R)

The earth's minimum velocity occurs when R = 1.520941962E+11 m (@ B)

v = (2 x 7.02944537126484E+22) ÷ (31558149 x 1.520941962E+11)

   = **29290.5355716777** m/s

29290.5355716777 m/s (calculated using; h=v.R)

The above confirms Kepler's 'swept-area = time' rule and shows that

v ∝ 1/R

or v = k/R

where k = 2.A / t

A-4.6.8 Centrifugal force in an orbiting body

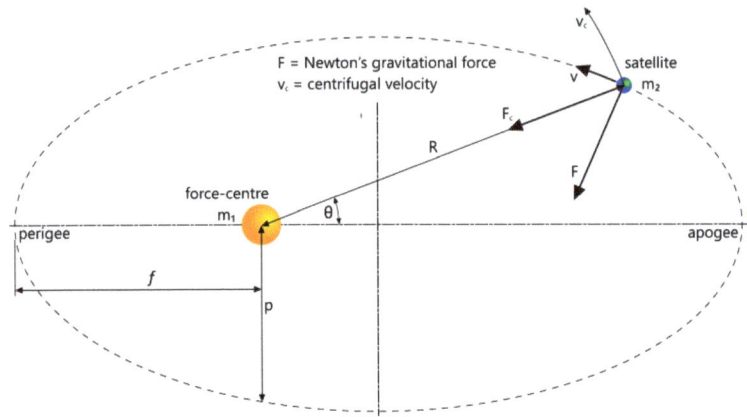

In any orbiting system, the centripetal force, i.e. Newton's *gravitational* force, must be equal to the orbiting body's centrifugal force, which may be calculated thus:

$F = m_1.v_1^2 / R$

$Fc = m_1.v_2^2 / R$

where

$v_2 = \sqrt{[G.m_1 / R]}$

@ the perihelion (perigee) of an ellipse; $Fc = F . f/p = F / (1+e)$

@ the aphelion (apogee) of an ellipse; $Fc = F . p/f = F . (1+e)$

Orbital velocity anywhere in an orbit may be calculated thus:

$v_2 = 2\pi.a.b / R.t$

where: t is the satellite's orbital period

Centrifugal velocity anywhere in an orbit may be calculated thus:

$\alpha = \sqrt{[^4/_3.\pi]}$

$\zeta = \sqrt{[(f.\mathrm{Sin}(\theta/2)^\alpha + p.\mathrm{Cos}(\theta/2)^\alpha) / (f.\mathrm{cos}(\theta/2)^\alpha + p.\mathrm{Sin}(\theta/2)^\alpha)]}$

$V_c = \zeta.v_2$

A-4.6.9 **Fundamental Laws of Orbital Motion**

1) Every orbital system must have a force-centre and at least one satellite

2) A force-centre's *mass* defines its satellite's orbital shapes and periods

3) Satellite orbits define a force-centre's spin

4) Sub-satellite orbits and force-centre spin define a satellite's spin

5) Sub-satellites have no effect on the force-centre

6) Satellites may be swapped between orbits without altering orbital shapes and periods; e.g. Jupiter may replace Earth and Jupiter would follow the same orbital path that Earth previously followed and would orbit in 365¼ days

www.ingramcontent.com/pod-product-compliance
Lightning Source LLC
Chambersburg PA
CBHW040322220526
45473CB00009B/2537